NEWPORT PUBLIC LIBRARY
W9-BSV-856
j920 Ford
Middleton, Haydn.
Henry Ford : the people's
NPT 31540001977600
DISCARD

MAR 0 1 2000

✷ WHAT'S THEIR STORY? ✷

Henry Ford

Oxford University Press, 198 Madison Avenue, New York, New York 10016

Oxford New York
Athens Auckland Bangkok Bogotá Bombay
Buenos Aires Calcutta Cape Town Dar es Salaam
Delhi Florence Hong Kong Istanbul Karachi
Kuala Lumpur Madras Madrid Melbourne
Mexico City Nairobi Paris Singapore
Taipei Tokyo Toronto Warsaw

and associated companies in
Berlin Ibadan

Oxford is a trademark of Oxford University Press

Text © Haydn Middleton 1997
Illustrations © Oxford University Press 1997

Originally published by Oxford University Press UK in 1997.

All rights reserved. No part of this publication may be reproduced, stored in a retrieval system, or transmitted, in any form or by any means, electronic, mechanical, photocopying, recording, or otherwise, without the prior permission of Oxford University Press.

ISBN 0-19-521406-4

1 3 5 7 9 10 8 6 4 2

Printed in Dubai by Oriental Press

✷ WHAT'S THEIR STORY? ✷

Henry Ford

THE PEOPLE'S CARMAKER

HAYDN MIDDLETON

Illustrated by Tony Morris

Newport Public Library

OXFORD UNIVERSITY PRESS

Henry Ford was born on a farm.
He grew up, more than a hundred years ago, in Michigan. He had five younger brothers and sisters, and they all had to work hard for their father. Much too hard, Henry thought. He wished they had more machines to help them in the fields.

Henry loved to play around with machines. If his brothers and sisters got wind-up toys for Christmas, they had to hide them because Henry always took them apart to see how they worked! "He's a tinkerer, not a farmer," his father used to say.

j920 Ford
Middleton, Haydn.
Henry Ford : the people's
NPT 31540001977600

3 1540 00197 7600

He was right. When Henry was 16, he walked to the nearby city of Detroit, and for the next three years he did odd jobs there, working with machines. Then he went back home. But his head was still full of machines. There was one machine in particular that he could not stop thinking about: the motor car.

A hundred years ago there were hardly any cars on the roads. In many parts of the world, there were very few real roads at all. If you wanted to get around, you had to use horses, boats, barges, or steam trains. But all these methods cost money. Ordinary working people could hardly ever afford to use them. As a result, they rarely traveled very far.

When he was in Detroit, Henry heard that the first motor cars were being made in Europe. Like today's cars, they used gasoline to make their engines go. People were amazed by them but thought they were very dangerous. In Britain someone had to walk in front of a slowly moving car, waving a red flag as a warning!

These first cars were so expensive that people called them "rich men's toys." But why, Henry wondered, should the rich have all the fun? Surely everybody should have one.

For several years Henry went on farming and tinkering. He planned one day to make cars—or "horseless carriages"—of his own. In the meantime, among other experiments, he built his own steam-powered tractor.

Henry did not work all the time. When he was 25 he married Clara Bryant, who had grown up on a nearby farm. She thought Henry was such a genius with machines that he was bound to succeed in the end. But he needed more money for his experiments. So he took a job in Detroit, as chief engineer at the electric power plant.

Henry and Clara bought a house in the city. Soon they had a son. They named him Edsel, after Henry's oldest friend. Henry worked hard at the power plant. But he worked even harder in his own backyard workshop, hammering together spare pieces of metal, until at last his first gasoline-driven car was ready for a test run.

It was almost two o'clock in the morning. Henry and his helper Jim Bishop rolled his strange-looking car toward the workshop doors.

Henry later called it his "baby carriage" because that is what it looked like. But there was a problem. He had made it too big to get through the doorway. So he simply snatched up a hammer and smashed a bigger hole in the brick wall. Then, with Clara watching, he started the car and drove off over the cobbled nighttime streets. Jim Bishop rode ahead on a bicycle, to make sure that no one got in the way. The car worked perfectly. The next day, Henry took Clara and Edsel for their first bumpy ride.

Henry's friends called him Crazy Henry because he was so mad about cars. When he was 36 he gave up his job and began to make and sell cars full-time. He built one car especially for racing: the 999.

When the 999 was ready, Henry wanted to test it right away, in a race against Alexander Winton's new car. But Clara stepped in. She was too scared to let Henry drive the racer. So Barney Oldfield, a bicycle rider, said he would do it. This was very brave. He had to learn to drive in only a week.

The day of the race came. Barney took an early lead and never let the other driver catch up. The noisy cars roared around the 5-mile (8-kilometer) course, and Barney finished in 5 minutes 28 seconds—an American record.

People now saw that Henry was not crazy at all. Soon rich businessmen were helping him to set up the Ford Motor Company. In its first five years the company produced eight different types, or models, of car. Each was named after a letter of the alphabet. The one that made Henry Ford's fortune was the Model T.

Henry wanted to build a motor car for the ordinary people. "It will be large enough for the family, but small enough for one person to run and care for," he said in 1907. "It will be made of the best materials, by the best workers, and in the simplest way possible. But it will be so low in price that any man who earns a good wage will be able to own one."

In 1908, his Model T cars went on sale. Costing $825 each, they were a terrific value for the money. They sold very quickly—and they kept on selling. The Model T was truly the People's Car. And the people loved it. They even gave it a nickname: the Tin Lizzie.

The Ford Motor Company made Tin Lizzies for 19 years. It sold 15 1/2 million of them in the United States, 1 million in Canada, and another quarter of a million in Britain. But it had taken Henry years to make his first car. How could he now make so many Model Ts?

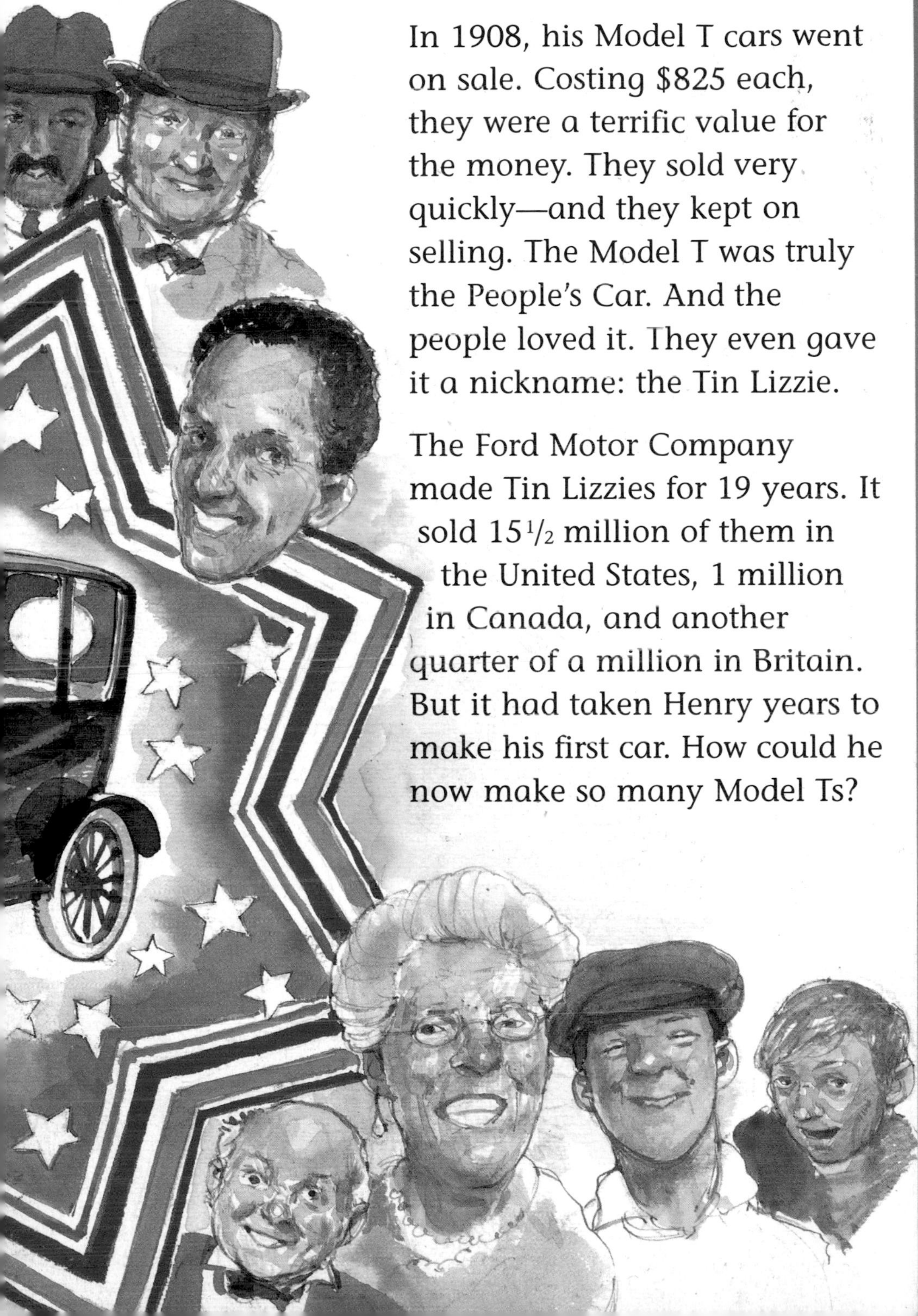

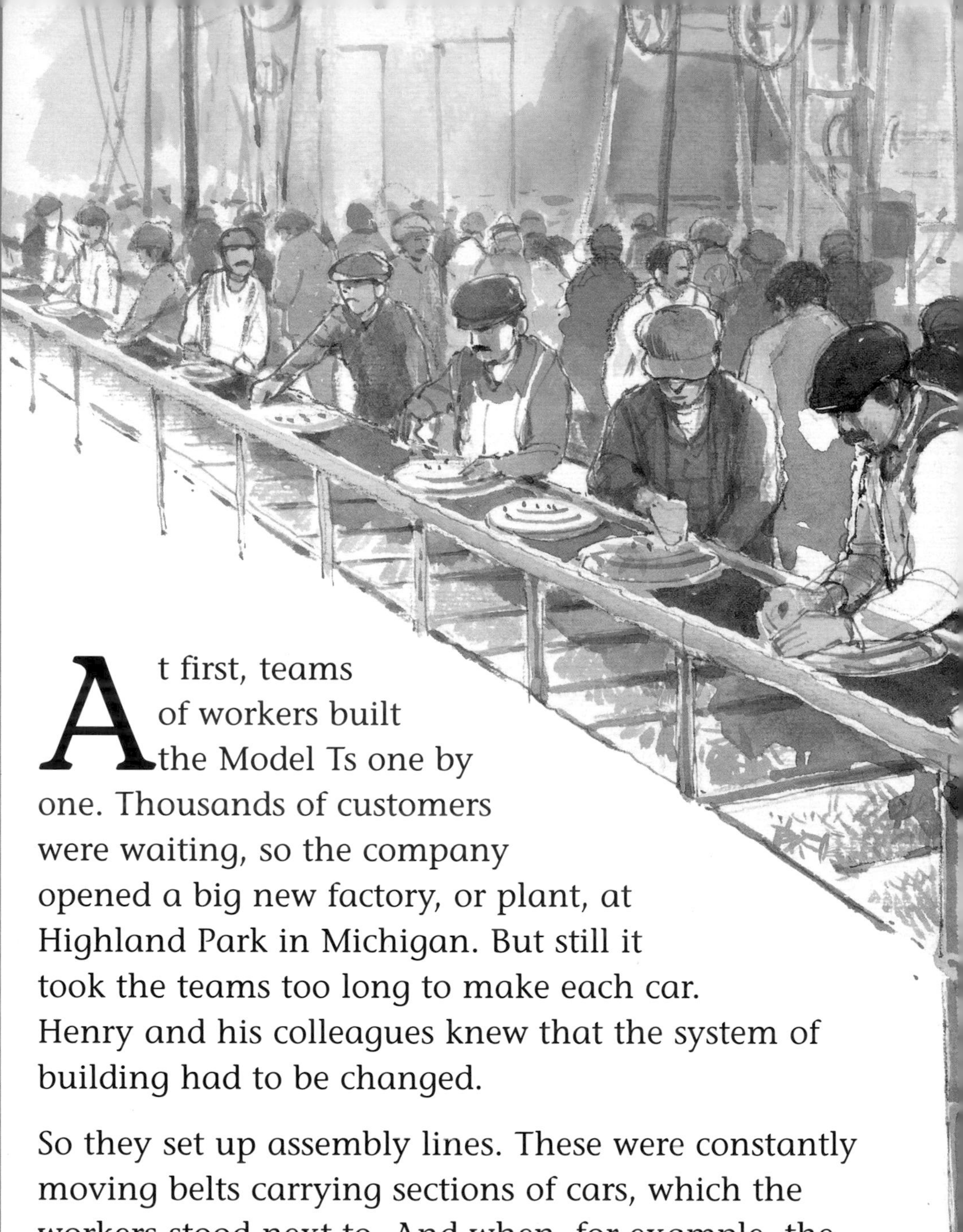

At first, teams of workers built the Model Ts one by one. Thousands of customers were waiting, so the company opened a big new factory, or plant, at Highland Park in Michigan. But still it took the teams too long to make each car. Henry and his colleagues knew that the system of building had to be changed.

So they set up assembly lines. These were constantly moving belts carrying sections of cars, which the workers stood next to. And when, for example, the cars' engines slowly passed by, the workers would fit pistons or crankshafts to them.

Each man had just one simple job to do. "The man who places a part does not fasten it," said Henry. "The man who puts in a bolt does not put on the nut; the man who puts on the nut does not tighten it." Cars could be made much faster this way, and then be sold even more cheaply. But for the workers, this new way of car making could be terribly boring.

Henry hit on a good plan to stop his workers on the assembly line from feeling bored. He doubled their wages from $2.50 to $5 a day. That same week, thousands of other men hurried to Highland Park to ask for jobs. A riot broke out, and the police had to soak everyone with fire hoses to stop it.

Henry chose his workers carefully. He paid them good wages, but they had to lead good, clean lives, both in the plant and at home. A special department was set up to make sure no one was misbehaving. Henry kept a close eye on the plant's top managers too.

The Ford Motor Company was like one big family, with Henry as its father. Some men found him much too bossy, so they left to take other jobs. But one man never had the chance to leave. This was Henry's real-life son, Edsel.

Henry was always very fond of Edsel. He taught the boy to drive when he was just eight. In those days there was no law against children driving. So Henry also gave him a car to drive himself to school and back.

When Edsel reached the age of 21, Henry took him to a bank in Detroit, showed him a million dollars in gold—then gave it to him as a birthday present! Edsel never did any job except work for his father. Sometimes that was hard. Henry rarely treated him like a grown-up, but he had everything else that he wanted. His father was, after all, one of the world's richest and most famous men.

For a while Henry even had to wear a false beard, because he was so tired of people recognizing him and stopping him in the street.

Like Clara in the early days, many people thought he was a genius. They thought he could do absolutely anything. Unfortunately, so did Henry now.

In 1914 the First World War broke out in Europe. Henry was disgusted. He thought ordinary people should be working and buying cars, not killing one another. So, with some friends, he decided he would end the war.

His plan was to go to Europe and keep talking to both sides until they agreed to stop fighting. Thousands of people came to watch his ocean liner set sail from New York. Some of them thought Henry really had gone crazy this time. One man sent him two squirrels in a cage. He wrote that they were "To go with the Nuts" on board the liner.

When the Peace Ship arrived in Norway, the newspaper reporters thought he was foolish, too. Soon Henry realized that he could not tell a whole continent what to do. Quietly he sailed home, where many people were still laughing at him. He had failed to make peace. Now he went back to what he did best: making cars.

The car plant at Highland Park was huge. But Henry had an even grander dream. And finally, after years of hard work, he made his dream come true.

In 1927 he completed a new superplant on the banks of the nearby Rouge River. It was the most gigantic assembly line ever seen. At one end all the raw materials poured in—iron ore, coal, timber, limestone, silica for glass. After just 28 hours, finished cars poured out at the other end. There were more than 90 different buildings for 42,000 workers. Five thousand men with mops and buckets did nothing all day long except keep the place clean.

Here the company built new types of cars: first the Model A, then the V-8. Henry still made only good, cheap cars and his company was now busy in 33 different countries. In his homeland people told each other stories about him as if he were a fairy-tale hero.

As Henry grew rich and famous, he made friends with other rich and famous people. These were men like Thomas Edison, the inventor of electric light; Harvey Firestone, the rubber-tire maker; and John Burroughs, a brilliant scientist. Every summer the four of them went on camping trips together.

One year, they stopped at a service station. While the attendant replaced a broken car headlight, Henry said to him, "The man who invented this light is over there."

"Thomas Edison!" gasped the attendant.

"Yes," said Henry, "and I am Henry Ford. And that man next to him is Harvey Firestone, who makes these tires."

Suddenly the attendant grew suspicious. He could not believe that so many well-known men were traveling together. Then he saw John Burroughs with his long white beard. "Look here, mister," he said to Henry, "if you're going to tell me that other guy is Santa Claus, I'm going to call the sheriff!"

When the Model T went on sale in 1908, Henry was 45 years old. As he got older, he found it harder to run the business. Times were tough all over the world in the 1930s. Henry could no longer afford to pay his workers so well. The Ford Motor Company was still a big family, but it was not a happy one.

Other companies were now building more luxurious, more modern cars. Edsel wanted Henry to do the same. But Henry was stubborn. He had always liked everything to be plain and simple. "The customer can have the Model T painted any color he wants," he once said, "so long as it is black."

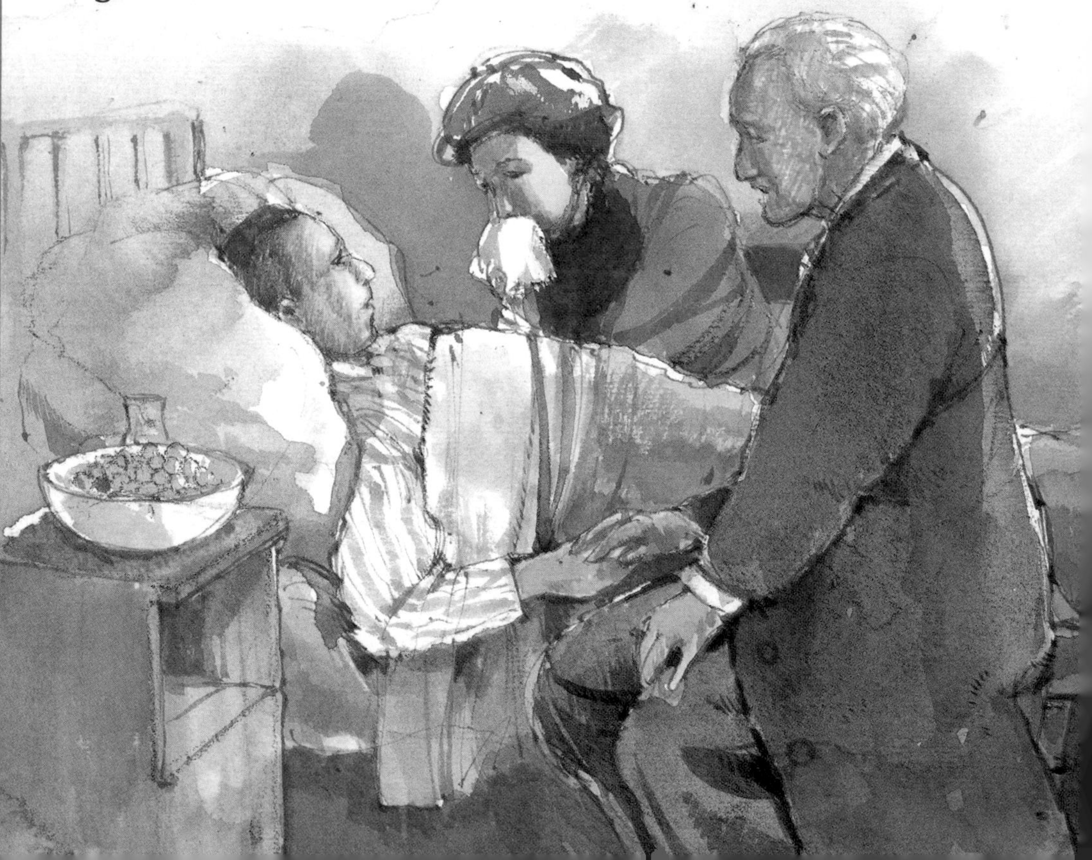

By the 1940s the company was in quite a mess. Edsel grew sick with worry and, at the age of 49, he died. Edsel's son, Henry Ford II, took over his father's job. One day he would make the company great again. But before then, old Henry himself died.

Henry had never been interested in making money to buy things. His goal, he said, was to make the world a little better.

He died in 1947 at age 83. Thanks mainly to him, car making had become a crucial industry in the United States. But by making cheap cars for ordinary people, Henry Ford had helped to make even greater changes. When he left his father's farm as a young man, four

Americans in every five lived cut off from one another on farms. When Henry died, four in every five lived in cities. And all these cities were now linked together by the finest system of roads on earth.

People remember that Henry Ford once said that "History is bunk." He also said that the present is far more important than the past. Yet few people have had a bigger effect on the history of the 20th century than Henry Ford.

Important dates in Henry Ford's life

1863 Born on family farm in Dearborn, Michigan.
1879 Walks to Detroit to work in machine shops.
1881 Returns to family farm.
1888 Marries Clara Bryant.
1893 Edsel Ford born.
1896 Completes his first gasoline-engined car, "the baby carriage."
1901 His 999 racing car sets US speed record.
1903 Sets up the Ford Motor Company.
1908 Production of Model T begins.
1914 Doubles workers' wages, from $2.50 to $5.00 a day.
1915 Travels to Europe on a liner to try to end First World War.
1918 Resigns as head of the company in favor of his son, Edsel.
1927 New Ford plant opened in Rouge River, Michigan.
1943 Edsel Ford dies.
1945 Grandson, Henry Ford II, takes over as head of the company.
1947 Dies at home, aged 83.

Index